PIERRE-SIMON LAPLACE

GREAT ASTRONOMERS

ROBERT STAWELL BALL

GIFT *Certificate*

TO:

FROM:

DATE:

PIERRE-SIMON LAPLACE

GREAT ASTRONOMERS

BY

ROBERT STAWELL BALL

D.Sc, LL.D., F.R.S.

(July 1, 1840 – November 25, 1913)

LOWNDEAN PROFESSOR OF ASTRONOMY AND GEOMETRY
IN THE UNIVERSITY OF CAMBRIDGE

DIAMOND BOOKS™

www.diamondbooks.ca

TORONTO, CANADA – 2017

DIAMOND BOOKS - CANADA

DIAMOND BOOKS

Toronto, ON, CANADA
http://www.diamondbooks.ca

BIBLIOGRAPHIC INFORMATION

'PIERRE-SIMON LAPLACE'
Is an extract or a chapter from the book
"GREAT ASTRONOMERS" by Robert Stawell Ball.

FIRST PUBLISHED IN: 1896.

PUBLISHED IN CANADA

Published in Canada by DIAMOND BOOKS - CANADA, an imprint of DIAMOND PUBLISHERS - http://www.diamondpublishers.com

FIRST EDITION: APRIL, 2017.

PAPERBACK EDITION : ISBN: 978-1-988357-60-7

PRINTED IN CANADA

PREFACE
TO THE ORIGINAL EDITION

It has been my object in these pages to present the life of each astronomer in such detail as to enable the reader to realise in some degree the man's character and surroundings; and I have endeavoured to indicate as clearly as circumstances would permit the main features of the discoveries by which he has become known.

There are many types of astronomers—from the stargazer who merely watches the heavens, to the abstract mathematician who merely works at his desk; it has, consequently, been necessary in the case of some lives to adopt a very different treatment from that which seemed suitable for others.

ROBERT STAWELL BALL.

The Observatory, Cambridge.
October, 1895.

NOTE: The full-length of 'Preface' can be found in its original edition of the book "GREAT ASTRONOMERS".

INTRODUCTION

Of all the natural sciences there is not one which offers such sublime objects to the attention of the inquirer as does the science of astronomy. From the earliest ages the study of the stars has exercised the same fascination as it possesses at the present day. Among the most primitive peoples, the movements of the sun, the moon, and the stars commanded attention from their supposed influence on human affairs. The practical utilities of astronomy were also obvious in primeval times. Maxims of extreme antiquity show how the avocations of the husbandman are to be guided by the movements of the heavenly bodies. The positions of the stars indicated the time to plough, and the time to sow. To the mariner who was seeking a way across the trackless ocean, the heavenly bodies .offered the only reliable marks by which his path could be guided. There was, accordingly, a stimulus both from intellectual curiosity and from practical necessity to follow the movements of the stars. Thus began a search for the causes of the ever-varying phenomena which the heavens display.

- ROBERT STAWELL BALL

Many of the earliest discoveries are indeed prehistoric. The great diurnal movement of the heavens, and the annual revolution of the sun, seem to have been known in times far more ancient than those to which any human monuments can be referred. The acuteness of the early observers enabled them to single out the more important of the wanderers which we now call planets. They saw that the star-like objects, Jupiter, Saturn, and Mars, with the more conspicuous Venus, constituted a class of bodies wholly distinct from the fixed stars among which their movements lay, and to which they bear such a superficial resemblance. But the penetration of the early astronomers went even further, for they recognized that Mercury also belongs to the same group, though this particular object is seen so rarely. It would seem that eclipses and other phenomena were observed at Babylon from a very remote period, while the most ancient records of celestial observations that we possess are to be found in the Chinese annals. The study of astronomy, in the sense in which we understand the word, may be said to have commenced under the reign of the Ptolemies at Alexandria. The most famous name in the science of this period is that of Hipparchus, who lived and worked at Rhodes about the year 160 B.C. It was his splendid investigations that first wrought the observed facts into a coherent branch of knowledge. He recognized the primary obligation which lies on the student of the heavens to compile as complete an inventory as possible of the objects which are there to be found. Hipparchus accordingly commenced by undertaking, on a small scale, a task exactly similar to that on which modern astronomers, with all available appliances of meridian circles, and photographic telescopes, are constantly engaged at the present day. He compiled a catalogue of the principal fixed stars, which is of special value to astronomers, as being the earliest work of its kind which has been handed down. He also studied the movements of the sun and the moon, and framed theories to account for the incessant changes which he saw in progress. He found a much more difficult problem in his attempt to interpret satisfactorily the complicated movements of the planets. With the view of constructing a theory which should give some coherent account of the subject, he made many observations of the places of these wandering stars. How great were the advances which Hipparchus accomplished may be appreciated if we reflect that, as a preliminary task to his more purely astronomical labours, he had to invent that branch of mathematical science by which alone the problems he proposed could be solved. It was for this purpose that he devised the indispensable method of calculation which we now know so well as trigonometry. Without the aid rendered by this beautiful art it would have been impossible for any n ally important advance in astronomical calculation to have been effected.

But the discovery which shows, beyond all others, that Hipparchus possessed one of the master-minds of all time was the detection of that remarkable celestial movement known as the precession of the equinoxes. The inquiry which conducted to this discovery involved a most profound investigation, especially when it is remembered that in the days of Hipparchus the means of observation of the heavenly bodies were only of the rudest description, and the available observations of earlier dates were extremely scanty. We can but look with astonishment on the genius of the man who, in spite of such difficulties, was able to detect such a phenomenon as the precession, and to exhibit its actual magnitude. I shall endeavour to explain the nature of this singular celestial movement, for it may be said to offer the first instance in the history of science in which we find that combination of accurate observation with skilful interpretation, of which, in the subsequent development of astronomy, we have so many splendid examples.

The word equinox implies the condition that the night is equal to the day. To a resident on the equator the night is no doubt equal to the day at all times in the year, but to one who lives on any other part of the earth, in either hemisphere, the night and the day are not generally equal. There is, however, one occasion in spring, and another in autumn, on which the day and the night are each twelve hours at all places on the earth. When the night and day are equal in spring, the point which the sun occupies on. the heavens is termed the vernal equinox. There is similarly another point in which the sun is situated at the time of the autumnal equinox. In any investigation of the celestial movements the positions of these two equinoxes on the heavens are of primary importance, and Hipparchus, with the instinct of genius, perceived their significance, and commenced to study them. It will be understood that we can always define the position of a point on the sky with reference to the surrounding stars. No doubt we do not see the stars near the sun when the sun is shining, but they are there nevertheless. The ingenuity of Hipparchus enabled him to determine the positions of each of the two equinoxes relatively to the stars which lie in its immediate vicinity. After examination of the celestial places of these points at different periods, he was led to the conclusion that each equinox was moving relatively to the stars, though that movement was so slow that twenty five thousand years would necessarily elapse before a complete circuit of the heavens was accomplished. Hipparchus traced out this phenomenon, and established it on an impregnable basis, so that all astronomers have ever since recognised the precession of the equinoxes) as one of the fundamental facts of astronomy. Not until nearly two thousand years after Hipparchus had made this splendid discovery was the explanation of its cause given by Newton.

From the days of Hipparchus down to the present hour the science of astronomy has steadily grown. One great observer after another has appeared from time to time, to reveal some new phenomenon with regard to the celestial bodies or their movements, while from time to time one commanding intellect after another has arisen to explain the true import of the facts of observations. The history of astronomy thus becomes inseparable from the history of the great men to whose labours its development is due. In the ensuing chapters we have endeavoured to sketch the lives and the work of the great philosophers, by whose labours the science of astronomy has been created.

We shall commence with Ptolemy, who, after the foundations of the science had been laid by Hipparchus, gave to astronomy the form in which it was taught throughout the Middle Ages. We shall next see the mighty revolution in our conceptions of the universe which are associated with the name of Copernicus. We then pass to those periods illumined by the genius of Galileo and Newton, and afterwards we shall trace the careers of other more recent discoverers, by whose industry and genius the boundaries of human knowledge have been so greatly extended. Our history will be brought down late enough to include some of the illustrious astronomers who laboured in the generation which has just passed away.

- ROBERT STAWELL BALL

TABLE OF CONTENTS

CHAPTERS

**An original Chapter from the book: "GREAT ASTRONOMERS", by Robert Stawell Ball.*

CHAPTER 11

GREAT ASTRONOMERS

Pierre-Simon Laplace

(March 23, 1749 – March 5, 1827)

Pierre-Simon Laplace
(March 23, 1749 – March 5, 1827)

CHAPTER 11

GREAT ASTRONOMERS: PIERRE-SIMON LAPLACE

THE author of the "Mecanique Celeste" was born at Beaumonten-Auge, near Honfleur, in 1749, just thirteen years later than his renowned friend Lagrange. His father was a farmer, but appears to have been in a position to provide a good education for a son who seemed promising. Considering the unorthodoxy in religious matters which is generally said to have characterized Laplace in later years, it is interesting to note that when he was a boy the subject which first claimed his attention was theology. He was, however, soon introduced to the study of mathematics, in which he presently became so proficient, that while he was still no more than eighteen years old, he obtained employment as a mathematical teacher in his native town.

Desiring wider opportunities for study and for the acquisition of fame than could be obtained in the narrow associations of provincial life, young Laplace started for Paris, being provided with letters of introduction to D'Alembert, who then occupied the most prominent position as a mathematician in France, if not in the whole of Europe. D'Alembert's fame was indeed so brilliant that Catherine the Great wrote to ask him to undertake the education of her son, and promised the splendid income of a hundred thousand francs. He preferred, however, a quiet life of research in Paris, although there was but a modest salary attached to his office. The philosopher accordingly declined the alluring offer to go to Russia, even though Catherine wrote again to say: "I know that your refusal arises from your desire to cultivate your studies and your friendships in quiet. But this is of no consequence: bring all your friends with you, and I promise you that both you and they shall have every accommodation in my power." With equal firmness the illustrious mathematician resisted the manifold attractions with which Frederick the Great sought to induce him to take up his residence at Berlin. In reading of these invitations we cannot but be struck at the extraordinary respect which was then paid to scientific distinction. It must be remembered that the discoveries of such a man as D'Alembert were utterly incapable of being appreciated except by those who possessed a high degree of mathematical culture. We nevertheless find the potentates of Russia and Prussia entreating and, as it happens, vainly entreating, the most distinguished mathematician in France to accept the positions that they were proud to offer him.

- ROBERT STAWELL BALL

It was to D'Alembert, the profound mathematician, that young Laplace, the son of the country farmer, presented his letters of introduction. But those letters seem to have elicited no reply, whereupon Laplace wrote to D'Alembert submitting a discussion of some point in Dynamics. This letter instantly produced the desired effect. D'Alembert thought that such mathematical talent as the young man displayed was in itself the best of introductions to his favour. It could not be overlooked, and accordingly he invited Laplace to come and see him. Laplace, of course, presented himself, and ere long D'Alembert obtained for the rising philosopher a professorship of mathematics in the Military School in Paris. This gave the brilliant young mathematician the opening for which he sought, and he quickly availed himself of it.

Laplace was twenty-three years old when his first memoir on a profound mathematical subject appeared in the Memoirs of the Academy at Turin. From this time onwards we find him publishing one memoir after another in which he attacks, and in many cases successfully vanquishes, profound difficulties in the application of the Newtonian theory of gravitation to the explanation of the solar system. Like his great contemporary Lagrange, he loftily attempted problems which demanded consummate analytical skill for their solution. The attention of the scientific world thus became riveted on the splendid discoveries which emanated from these two men, each gifted with extraordinary genius.

Laplace's most famous work is, of course, the “Mecanique Celeste,” in which he essayed a comprehensive attempt to carry out the principles which Newton had laid down, into much greater detail than Newton had found practicable. The fact was that Newton had not only to construct the theory of gravitation, but he had to invent the mathematical tools, so to speak, by which his theory could be applied to the explanation of the movements of the heavenly bodies. In the course of the century which had elapsed between the time of Newton and the time of Laplace, mathematics had been extensively developed. In particular, that potent instrument called the infinitesimal calculus, which Newton had invented for the investigation of nature, had become so far perfected that Laplace, when he attempted to unravel the movements of the heavenly bodies, found himself provided with a calculus far more efficient than that which had been available to Newton. The purely geometrical methods which Newton employed, though they are admirably adapted for demonstrating in a general way the tendencies of forces and for explaining the more obvious phenomena by which the movements of the heavenly bodies are disturbed, are yet quite inadequate for dealing with the more subtle effects of the Law of Gravitation.

The disturbances which one planet exercises upon the rest can only be fully ascertained by the aid of long calculation, and for these calculations analytical methods are required.

With an armament of mathematical methods which had been perfected since the days of Newton by the labours of two or three generations of consummate mathematical inventors, Laplace essayed in the "Mecanique Celeste" to unravel the mysteries of the heavens. It will hardly be disputed that the book which he has produced is one of the most difficult books to understand that has ever been written. In great part, of course, this difficulty arises from the very nature of the subject, and is so far unavoidable. No one need attempt to read the "Mecanique Celeste" who has not been naturally endowed with considerable mathematical aptitude which he has cultivated by years of assiduous study.

- ROBERT STAWELL BALL

The critic will also note that there are grave defects in Laplace's method of treatment. The style is often extremely obscure, and the author frequently leaves great gaps in his argument, to the sad discomfiture of his reader. Nor does it mend matters to say, as Laplace often does say, that it is "easy to see" how one step follows from another. Such inferences often present great difficulties even to excellent mathematicians. Tradition indeed tells us that when Laplace had occasion to refer to his own book, it sometimes happened that an argument which he had dismissed with his usual formula, "Il est facile a voir," cost the illustrious author himself an hour or two of hard thinking before he could recover the train of reasoning which had been omitted. But there are certain parts of this great work which have always received the enthusiastic admiration of mathematicians. Laplace has, in fact, created whole tracts of science, some of which have been subsequently developed with much advantage in the prosecution of the study of Nature.

Judged by a modern code, the gravest defect of Laplace's great work is rather of a moral than of a mathematical nature. Lagrange and he advanced together in their study of the mechanics of the heavens, at one time perhaps along parallel lines, while at other times they pursued the same problem by almost identical methods. Sometimes the important result was first reached by Lagrange, sometimes it was Laplace who had the good fortune to make the discovery. It would doubtless be a difficult matter to draw the line which should exactly separate the contributions to astronomy made by one of these illustrious mathematicians, and the contributions made by the other. But in his great work Laplace in the loftiest manner disdained to accord more than the very barest recognition to Lagrange, or to any of the other mathematicians, Newton alone excepted, who had advanced our knowledge of the mechanism of the heavens. It would be quite impossible for a student who confined his reading to the "Mecanique Celeste" to gather from any indications that it contains whether the discoveries about which he was reading had been really made by Laplace himself or whether they had not been made by Lagrange, or by Euler, or by Clairaut. With our present standard of morality in such matters, any scientific man who now brought forth a work in which he presumed to ignore in this wholesale fashion the contributions of others to the subject on which he was writing, would be justly censured and bitter controversies would undoubtedly arise. Perhaps we ought not to judge Laplace by the standard of our own time, and in any case I do not doubt that Laplace might have made a plausible defence. It is well known that when two investigators are working at the same subjects, and constantly publishing their results, it sometimes becomes difficult for each investigator himself to distinguish exactly between what he has accomplished and that which must be credited to his rival. Laplace may probably have said to himself that he was going to devote his energies to a great work on the interpretation of Nature, that it would take all his time and all his faculties, and all the resources of knowledge that he could command, to deal justly with the mighty problems before him. He would not allow himself to be distracted by any side issue. He could not tolerate that pages should be wasted in merely discussing to whom we owe each formula, and to whom each deduction from such formula is due. He would rather endeavour to produce as complete a picture as he possibly could of the celestial mechanics, and whether it were by means of his mathematics alone, or whether the discoveries of others may have contributed in any degree to the result, is a matter so infinitesimally insignificant in comparison with the grandeur of his subject that he would altogether neglect it.

- ROBERT STAWELL BALL

Laplace might say, “If Lagrange should think that his discoveries had been unduly appropriated, the proper course would be for him to do exactly what I have done. Let him also write a ‘Mecanique Celeste,’ let him employ those consummate talents which he possesses in developing his noble subject to the utmost. Let him utilise every result that I or any other mathematician have arrived at, but not trouble himself unduly with unimportant historical details as to who discovered this, and who discovered that; let him produce such a work as he could write, and I shall heartily welcome it as a splendid contribution to our science.”

Certain it is that Laplace and Lagrange continued the best of friends, and on the death of the latter it was Laplace who was summoned to deliver the funeral oration at the grave of his great rival. The investigations of Laplace are, generally speaking, of too technical a character to make it possible to set forth any account of them in such a work as the present. He did publish, however, one treatise, called the "Systeme du Monde," in which, without introducing mathematical symbols, he was able to give a general account of the theories of the celestial movements, and of the discoveries to which he and others had been led. In this work the great French astronomer sketched for the first time that remarkable doctrine by which his name is probably most generally known to those readers of astronomical books who are not specially mathematicians. It is in the "Systeme du Monde" that Laplace laid down the principles of the Nebular Theory which, in modern days, has been generally accepted by those philosophers who are competent to judge, as substantially a correct expression of a great historical fact.

The Nebular Theory gives a physical account of the origin of the solar system, consisting of the sun in the centre, with the planets and their attendant satellites. Laplace perceived the significance of the fact that all the planets revolved in the same direction around the sun; he noticed also that the movements of rotation of the planets on their axes were performed in the same direction as that in which a planet revolves around the sun; he saw that the orbits of the satellites, so far at least as he knew them, revolved around their primaries also in the same direction. Nor did it escape his attention that the sun itself rotated on its axis in the same sense. His philosophical mind was led to reflect that such a remarkable unanimity in the direction of the movements in the solar system demanded some special explanation. It would have been in the highest degree improbable that there should have been this unanimity unless there had been some physical reason to account for it.

To appreciate the argument let us first concentrate our attention on three particular bodies, namely the earth, the sun, and the moon. First the earth revolves around the sun in a certain direction, and the earth also rotates on its axis. The direction in which the earth turns in accordance with this latter movement might have been that in which it revolves around the sun, or it might of course have been opposite thereto. As a matter of fact the two agree. The moon in its monthly revolution around the earth follows also the same direction, and our satellite rotates on its axis in the same period as its monthly revolution, but in doing so is again observing this same law. We have therefore in the earth and moon four movements, all taking place in the same direction, and this is also identical with that in which the sun rotates once every twenty-five days. Such a coincidence would be very unlikely unless there were some physical reason for it. Just as unlikely would it be that in tossing a coin five heads or five tails should follow each other consecutively. If we toss a coin five times the chances that, it will turn up all heads or all tails is but a small one. The probability of such an event is only one-sixteenth.

There are, however, in the solar system many other bodies besides the three just mentioned which are animated by this common movement. Among them are, of course, the great planets, Jupiter, Saturn, Mars, Venus, and Mercury, and the satellites which attend on these planets. All these planets rotate on their axes in the same direction as they revolve around the sun, and all their satellites revolve also in the same way. Confining our attention merely to the earth, the sun, and the five great planets with which Laplace was acquainted, we have no fewer than six motions of revolution and seven motions of rotation, for in the latter we include the rotation of the sun. We have also sixteen satellites of the planets mentioned whose revolutions round their primaries are in the same direction. The rotation of the moon on its axis may also be reckoned, but as to the rotations of the satellites of the other planets we cannot speak with any confidence, as they are too far off to be observed with the necessary accuracy. We have thus thirty circular movements in the solar system connected with the sun and moon and those great planets than which no others were known in the days of Laplace. The significant fact is that all these thirty movements take place in the same direction. That this should be the case without some physical reason would be just as unlikely as that in tossing a coin thirty times it should turn up all heads or all tails every time without exception.

We can express the argument numerically. Calculation proves that such an event would not generally happen oftener than once out of five hundred millions of trials. To a philosopher of Laplace's penetration, who had made a special study of the theory of probabilities, it seemed well-nigh inconceivable that there should have been such unanimity in the celestial movements, unless there had been some adequate reason to account for it. We might, indeed, add that if we were to include all the objects which are now known to belong to the solar system, the argument from probability might be enormously increased in strength. To Laplace the argument appeared so conclusive that he sought for some physical cause of the remarkable phenomenon which the solar system presented. Thus it was that the famous Nebular Hypothesis took its rise. Laplace devised a scheme for the origin of the sun and the planetary system, in which it would be a necessary consequence that all the movements should take place in the same direction as they are actually observed to do.

Let us suppose that in the beginning there was a gigantic mass of nebulous material, so highly heated that the iron and other substances which now enter into the composition of the earth and planets were then suspended in a state of vapour.

There is nothing unreasonable in such a supposition; indeed, we know as a matter of fact that there are thousands of such nebulae to be discerned at present through our telescopes. It would be extremely unlikely that any object could exist without possessing some motion of rotation; we may in fact assert that for rotation to be entirely absent from the great primaeval nebula would be almost infinitely improbable. As ages rolled on, the nebula gradually dispersed away by radiation its original stores of heat, and, in accordance with well-known physical principles, the materials of which it was formed would tend to coalesce. The greater part of those materials would become concentrated in a mighty mass surrounded by outlying uncondensed vapours. There would, however, also be regions throughout the extent of the nebula, in which subsidiary centres of condensation would be found. In its long course of cooling, the nebula would, therefore, tend ultimately to form a mighty central body with a number of smaller bodies disposed around it. As the nebula was initially endowed with a movement of rotation, the central mass into which it had chiefly condensed would also revolve, and the subsidiary bodies would be animated by movements of revolution around the central body. These movements would be all pursued in one common direction, and it follows, from well-known mechanical principles, that each of the subsidiary masses, besides participating in the general revolution around the central body, would also possess a rotation around its axis, which must likewise be performed in the same direction. Around the subsidiary bodies other objects still smaller would be formed, just as they themselves were formed relatively to the great central mass, as the ages sped by, and the heat of these bodies became gradually dissipated, the various objects would coalesce, first into molten liquid masses, and thence, at a further stage of cooling, they would assume the appearance of solid masses, thus producing the planetary bodies such as we now know them. The great central mass, on account of its preponderating dimensions, would still retain, for further uncounted ages, a large quantity of its primeval heat, and would thus display the splendours of a glowing sun. In this way Laplace was able to account for the remarkable phenomena presented in the movements of the bodies of the solar system. There are many other points also in which the nebular theory is known to tally with the facts of observation. In fact, each advance in science only seems to make it more certain that the Nebular Hypothesis substantially represents the way in which our solar system has grown to its present form.

Not satisfied with a career which should be merely scientific, Laplace sought to connect himself with public affairs. Napoleon appreciated his genius, and desired to enlist him in the service of the State. Accordingly he appointed Laplace to be Minister of the Interior. The experiment was not successful, for he was not by nature a statesman. Napoleon was much disappointed at the ineptitude which the great mathematician showed for official life, and, in despair of Laplace's capacity as an administrator, declared that he carried the spirit of his infinitesimal calculus into the management of business. Indeed, Laplace's political conduct hardly admits of much defence. While he accepted the honours which Napoleon showered on him in the time of his prosperity, he seems to have forgotten all this when Napoleon could no longer render him service. Laplace was made a Marquis by Louis XVIII., a rank which he transmitted to his son, who was born in 1789. During the latter part of his life the philosopher lived in a retired country place at Arcueile. Here he pursued his studies, and by strict abstemiousness, preserved himself from many of the infirmities of old age.

He died on March the 5th, 1827, in his seventy-eighth year, his last words being, **"What we know is but little, what we do not know is immense."**

- ROBERT STAWELL BALL

COMPLIMENTARY MATERIAL

A BRIEF BIOGRAPHICAL SKETCH

PIERRE-SIMON LAPLACE

- ROBERT STAWELL BALL

Pierre-Simon Laplace
(March 23, 1749 – March 5, 1827)

COMPLIMENTARY MATERIAL

A BRIEF BIOGRAPHICAL SKETCH

PIERRE-SIMON LAPLACE

Pierre-Simon, marquis de Laplace (23 March 1749 – 5 March 1827) was an influential French scholar whose work was important to the development of mathematics, statistics, physics and astronomy. He summarized and extended the work of his predecessors in his five-volume Mécanique Céleste

(Celestial Mechanics) (1799–1825). This work translated the geometric study of classical mechanics to one based on calculus, opening up a broader range of problems. In statistics, the Bayesian interpretation of probability was developed mainly by Laplace.

Laplace formulated Laplace's equation, and pioneered the Laplace transform which appears in many branches of mathematical physics, a field that he took a leading role in forming. The Laplacian differential operator, widely used in mathematics, is also named after him. He restated and developed the nebular hypothesis of the origin of the Solar System and was one of the first scientists to postulate the existence of black holes and the notion of gravitational collapse.

Laplace is remembered as one of the greatest scientists of all time. Sometimes referred to as the French Newton or Newton of France, he has been described as possessing a phenomenal natural mathematical faculty superior to that of any of his contemporaries.

Early Years

The original documents relating to the life of Laplace were lost when the family château of Saint-Julien de Mailloc, near Lisieux, the home of his great-great-grandson the Comte de Colbert-Laplace burned in 1925 and some had been destroyed earlier, when his house at Arcueil near Paris was looted by house breakers in 1871. Laplace was born in Beaumont-en-Auge, Normandy on 23 March 1749, a village four miles west of Pont l'Eveque in Normandy. According to W. W. Rouse Ball, His father, Pierre de Laplace, owned and farmed the small estates of Maarquis. His great-uncle, Maitre Oliver de Laplace, had held the title of Chirurgien Royal. It would seem that from a pupil he became an usher in the school at Beaumont; but, having procured a letter of introduction to d'Alembert, he went to Paris to advance his fortune. However, Karl Pearson is scathing about the inaccuracies in Rouse

Ball's account and states:

Indeed Caen was probably in Laplace's day the most intellectually active of all the towns of Normandy. It was here that Laplace was educated and was provisionally a professor. It was here he wrote his first paper published in the Mélanges of the Royal Society of Turin, Tome iv. 1766–1769, at least two years before he went at 22 or 23 to Paris in 1771. Thus before he was 20 he was in touch with Lagrange in Turin. He did not go to Paris a raw self-taught country lad with only a peasant background! In 1765 at the age of sixteen Laplace left the "School of the Duke of Orleans" in Beaumont and went to the University of Caen, where he appears to have studied for five years and was a member of the Sphinx. The 'École Militaire' of Beaumont did not replace the old school until 1776.

His parents were from comfortable families. His father was Pierre Laplace, and his mother was Marie-Anne Sochon. The Laplace family was involved in agriculture until at least 1750, but Pierre

Laplace senior was also a cider merchant and syndic of the town of Beaumont.

Pierre Simon Laplace attended a school in the village run at a Benedictine priory, his father intending that he be ordained in the Roman Catholic Church. At sixteen, to further his father's intention, he was sent to the University of Caen to read theology.

At the university, he was mentored by two enthusiastic teachers of mathematics, Christophe Gadbled and Pierre Le Canu, who awoke his zeal for the subject. Here Laplace's brilliance as a mathematician was quickly recognised and while still at Caen he wrote a memoir Sur le Calcul integral aux differences infiniment petites et aux differences finies. This provided the first intercourse between Laplace and Lagrange for Lagrange who was the senior by thirteen years , had recently founded in his native city of Turin a journal named Miscellanea Taurinensia, in which many of his other early works were printed and it was in the fourth volume of this series the Laplace's paper appeared. About this time, recognizing that he had no vocation for the priesthood, he determined to become a professional mathematician. In this connection reference may perhaps be made to the statement, which has appeared in some notices of him, that he broke altogether with the church and became an atheist. Laplace did not graduate in theology but left for Paris with a letter of introduction from Le Canu to Jean le Rond d'Alembert who at that time was supreme in scientific circles.

According to his great-great-grandson, d'Alembert received him rather poorly, and to get rid of him gave him a thick mathematics book, saying to come back when he had read it. When Laplace came back a few days later, d'Alembert was even less friendly and did not hide his opinion that it was impossible that Laplace could have read and understood the book. But upon questioning him, he realized that it was true, and from that time he took Laplace under his care.

Another version is that Laplace solved overnight a problem that d'Alembert set him for submission the following week, then solved a harder problem the following night. D'Alembert was impressed and recommended him for a teaching place in the École Militaire.

With a secure income and undemanding teaching, Laplace now threw himself into original research and in the next seventeen years, 1771–1787, he produced much of his original work in astronomy.

Laplace further impressed the Marquis de Condorcet, and already in 1771 Laplace felt that he was entitled to membership of the French Academy of Sciences. However, in that year, admission went to Alexandre-Théophile Vandermonde and in 1772 to Jacques Antoine Joseph Cousin. Laplace was disgruntled, and at the beginning of 1773, d'Alembert wrote to Lagrange in Berlin to ask if a position could be found for Laplace there. However, Condorcet became permanent secretary of the Académie in February and Laplace was elected associate member on 31 March, at age 24.

On 15 March 1788, at the age of thirty-nine, Laplace married Marie-Charlotte de Courty de Romanges, a pretty eighteen-and-a-half-year-old girl from a good family in Besançon. The wedding was celebrated at Saint-Sulpice, Paris. The couple had a son, Charles-Émile (1789–1874), and a daughter, Sophie-Suzanne (1792–1813).

Analysis, Probability, and Astronomical Stability

Laplace's early published work in 1771 started with differential equations and finite differences but he was already starting to think about the mathematical and philosophical concepts of probability and statistics. However, before his election to the Académie in 1773, he had already drafted two papers that would establish his reputation. The first, Mémoire sur la probabilité des causes par les événements was ultimately published in 1774 while the second paper, published in 1776, further elaborated his statistical thinking and also began his systematic work on celestial mechanics and the stability of the Solar System. The two disciplines would always be interlinked in his mind. "Laplace took probability as an instrument for repairing defects in knowledge." Laplace's work on probability and statistics is discussed below with his mature work on the analytic theory of probabilities.

- ROBERT STAWELL BALL

Stability of the Solar System

Sir Isaac Newton had published his Philosophiae Naturalis Principia Mathematica in 1687 in which he gave a derivation of Kepler's laws, which describe the motion of the planets, from his laws of motion and his law of universal gravitation. However, though Newton had privately developed the methods of calculus, all his published work used cumbersome geometric reasoning, unsuitable to account for the more subtle higher-order effects of interactions between the planets. Newton himself had doubted the possibility of a mathematical solution to the whole, even concluding that periodic divine intervention was necessary to guarantee the stability of the Solar System. Dispensing with the hypothesis of divine intervention would be a major activity of Laplace's scientific life. It is now generally regarded that Laplace's methods on their own, though vital to the development of the theory, are not sufficiently precise to demonstrate the stability of the Solar System, and indeed, the Solar System is understood to be chaotic, although it happens to be fairly stable.

One particular problem from observational astronomy was the apparent instability whereby Jupiter's orbit appeared to be shrinking while that of Saturn was expanding. The problem had been tackled by Leonhard Euler in 1748 and Joseph Louis Lagrange in 1763 but without success. In 1776, Laplace published a memoir in which he first explored the possible influences of a purported luminiferous ether or of a law of gravitation that did not act instantaneously. He ultimately returned to an intellectual investment in Newtonian gravity. Euler and Lagrange had made a practical approximation by ignoring small terms in the equations of motion. Laplace noted that though the terms themselves were small, when integrated over time they could become important. Laplace carried his analysis into the higher-order terms, up to and including the cubic. Using this more exact analysis, Laplace concluded that any two planets and the sun must be in mutual equilibrium and thereby launched his work on the stability of the Solar System. Gerald James Whitrow described the achievement as "the most important advance in physical astronomy since Newton".

Laplace had a wide knowledge of all sciences and dominated all discussions in the Académie. Laplace seems to have regarded analysis merely as a means of attacking physical problems, though the ability with which he invented the necessary analysis is almost phenomenal. As long as his results were true he took but little trouble to explain the steps by which he arrived at them; he never studied elegance or symmetry in his processes, and it was sufficient for him if he could by any means solve the particular question he was discussing.

On the figure of the Earth

During the years 1784–1787 he published some memoirs of exceptional power. Prominent among these is one read in 1783, reprinted as Part II of Théorie du Mouvement et de la figure elliptique des planètes in 1784, and in the third volume of the Mécanique céleste. In this work, Laplace completely determined the attraction of a spheroid on a particle outside it. This is memorable for the introduction into analysis of spherical harmonics or Laplace's coefficients, and also for the development of the use of what we would now call the gravitational potential in celestial mechanics.

- ROBERT STAWELL BALL

Spherical Harmonics

In 1783, in a paper sent to the Académie, Adrien-Marie Legendre had introduced what are now known as associated Legendre functions. If two points in a plane have polar co-ordinates (r, θ) and (r ', θ'), where r ' ≥ r, then, by elementary manipulation, the reciprocal of the distance between the points, d, can be written as:

$$\frac{1}{d} = \frac{1}{r'}\left[1 - 2\cos(\theta' - \theta)\frac{r}{r'} + \left(\frac{r}{r'}\right)^2\right]^{-\frac{1}{2}}.$$

This expression can be expanded in powers of r/r ' using Newton's generalised binomial theorem to give:

$$\frac{1}{d} = \frac{1}{r'}\sum_{k=0}^{\infty} P_k^0(\cos(\theta' - \theta))\left(\frac{r}{r'}\right)^k.$$

The sequence of functions P0k(cosф) is the set of so-called "associated Legendre functions" and their usefulness arises from the fact that every function of the points on a circle can be expanded as a series of them.

Laplace, with scant regard for credit to Legendre, made the non-trivial extension of the result to three dimensions to yield a more general set of functions, the spherical harmonics or Laplace coefficients. The latter term is not in common use now .

- ROBERT STAWELL BALL

Potential Theory

This paper is also remarkable for the development of the idea of the scalar potential. The gravitational force acting on a body is, in modern language, a vector, having magnitude and direction. A potential function is a scalar function that defines how the vectors will behave. A scalar function is computationally and conceptually easier to deal with than a vector function.

Alexis Clairaut had first suggested the idea in 1743 while working on a similar problem though he was using Newtonian-type geometric reasoning. Laplace described Clairaut's work as being "in the class of the most beautiful mathematical productions". However, Rouse Ball alleges that the idea "was appropriated from Joseph Louis Lagrange, who had used it in his memoirs of 1773, 1777 and 1780". The term "potential" itself was due to Daniel Bernoulli, who introduced it in his 1738 memoire Hydrodynamica. However, according to Rouse Ball, the term "potential function" was not actually used (to refer to a function V of the coordinates of space in Laplace's sense) until George Green's 1828 An Essay on the Application of Mathematical Analysis to the Theories of Electricity and Magnetism.

Laplace applied the language of calculus to the potential function and showed that it always satisfies the differential equation:

$$\nabla^2 V = \frac{\partial^2 V}{\partial x^2} + \frac{\partial^2 V}{\partial y^2} + \frac{\partial^2 V}{\partial z^2} = 0.$$

An analogous result for the velocity potential of a fluid had been obtained some years previously by Leonhard Euler.

Laplace's subsequent work on gravitational attraction was based on this result. The quantity ∇2V has been termed the concentration of V and its value at any point indicates the "excess" of the value of V there over its mean value in the neighbourhood of the point. Laplace's equation, a special case of Poisson's equation, appears ubiquitously in mathematical physics. The concept of a potential occurs in fluid dynamics, electromagnetism and other areas. Rouse Ball speculated that it might be seen as "the outward sign" of one of the a priori forms in Kant's theory of perception.

The spherical harmonics turn out to be critical to practical solutions of Laplace's equation. Laplace's equation in spherical coordinates, such as are used for mapping the sky, can be simplified, using the method of separation of variables into a radial part, depending solely on distance from the centre point, and an angular or spherical part. The solution to the spherical part of the equation can be expressed as a series of Laplace's spherical harmonics, simplifying practical computation.

Planetary and Lunar Inequalities

Jupiter–Saturn great inequality

Laplace presented a memoir on planetary inequalities in three sections, in 1784, 1785, and 1786. This dealt mainly with the identification and explanation of the perturbations now known as the "great Jupiter–Saturn inequality". Laplace solved a longstanding problem in the study and prediction of the movements of these planets. He showed by general considerations, first, that the mutual action of two planets could never cause large changes in the eccentricities and inclinations of their orbits; but then, even more importantly, that peculiarities arose in the Jupiter–Saturn system because of the near approach to commensurability of the mean motions of Jupiter and Saturn.

In this context commensurability means that the ratio of the two planets' mean motions is very nearly equal to a ratio between a pair of small whole numbers. Two periods of Saturn's orbit around the Sun almost equal five of Jupiter's. The corresponding difference between multiples of the mean motions, (2nJ − 5nS), corresponds to a period of nearly 900 years, and it occurs as a small divisor in the integration of a very small perturbing force with this same period. As a result, the integrated perturbations with this period are disproportionately large, about 0.8° degrees of arc in orbital longitude for Saturn and about 0.3° for Jupiter.

Further developments of these theorems on planetary motion were given in his two memoirs of 1788 and 1789, but with the aid of Laplace's discoveries, the tables of the motions of Jupiter and Saturn could at last be made much more accurate. It was on the basis of Laplace's theory that Delambre computed his astronomical tables.

Books

Laplace now set himself the task to write a work which should "offer a complete solution of the great mechanical problem presented by the Solar System, and bring theory to coincide so closely with observation that empirical equations should no longer find a place in astronomical tables." The result is embodied in the Exposition du système du monde and the Mécanique céleste.

The former was published in 1796, and gives a general explanation of the phenomena, but omits all details. It contains a summary of the history of astronomy. This summary procured for its author the honour of admission to the forty of the French Academy and is commonly esteemed one of the masterpieces of French literature, though it is not altogether reliable for the later periods of which it treats.

Laplace developed the nebular hypothesis of the formation of the Solar System, first suggested by Emanuel Swedenborg and expanded by Immanuel Kant, a hypothesis that continues to dominate accounts of the origin of planetary systems. According to Laplace's description of the hypothesis, the Solar System had evolved from a globular mass of incandescent gas rotating around an axis through its centre of mass. As it cooled, this mass contracted, and successive rings broke off from its outer edge. These rings in their turn cooled, and finally condensed into the planets, while the sun represented the central core which was still left. On this view, Laplace predicted that the more distant planets would be older than those nearer the sun.

As mentioned, the idea of the nebular hypothesis had been outlined by Immanuel Kant in 1755, and he had also suggested "meteoric aggregations" and tidal friction as causes affecting the formation of the Solar System. Laplace was probably aware of this, but, like many writers of his time, he generally did not reference the work of others.

Laplace's analytical discussion of the Solar System is given in his Mécanique céleste published in five volumes. The first two volumes, published in 1799, contain methods for calculating the motions of the planets, determining their figures, and resolving tidal problems. The third and fourth volumes, published in 1802 and 1805, contain applications of these methods, and several astronomical tables. The fifth volume, published in 1825, is mainly historical, but it gives as appendices the results of Laplace's latest researches. Laplace's own investigations embodied in it are so numerous and valuable that it is regrettable to have to add that many results are appropriated from other writers with scanty or no acknowledgement, and the conclusions – which have been described as the organized result of a century of patient toil – are frequently mentioned as if they were due to Laplace.

Jean-Baptiste Biot, who assisted Laplace in revising it for the press, says that Laplace himself was frequently unable to recover the details in the chain of reasoning, and, if satisfied that the conclusions were correct, he was content to insert the constantly recurring formula, "Il est aisé à voir que..." ("It is easy to see that..."). The Mécanique céleste is not only the translation of Newton's Principia into the language of the differential calculus, but it completes parts of which Newton had been unable to fill in the details. The work was carried forward in a more finely tuned form in Félix Tisserand's Traité de mécanique céleste (1889–1896), but Laplace's treatise will always remain a standard authority.

Black Holes

Laplace also came close to propounding the concept of the black hole. He suggested that there could be massive stars whose gravity is so great that not even light could escape from their surface (see escape velocity).

- ROBERT STAWELL BALL

Arcueil

Laplace's house at Arcueil.

In 1806, Laplace bought a house in Arcueil, then a village and not yet absorbed into the Paris conurbation. Claude Louis Berthollet was a neighbour—their gardens were not separated—and the pair formed the nucleus of an informal scientific circle, latterly known as the Society of Arcueil. Because of their closeness to Napoleon, Laplace and Berthollet effectively controlled advancement in the scientific establishment and admission to the more prestigious offices. The Society built up a complex pyramid of patronage. In 1806, Laplace was also elected a foreign member of the Royal Swedish Academy of Sciences.

Analytic Theory of Probabilities

In 1812, Laplace issued his Théorie analytique des probabilités in which he laid down many fundamental results in statistics. The first half of this treatise was concerned with probability methods and problems, the second half with statistical methods and applications. Laplace's proofs are not always rigorous according to the standards of a later day, and his perspective slides back and forth between the Bayesian and non-Bayesian views with an ease that makes some of his investigations difficult to follow, but his conclusions remain basically sound even in those few situations where his analysis goes astray. In 1819, he published a popular account of his work on probability. This book bears the same relation to the Théorie des probabilités that the Système du monde does to the Méchanique céleste.

Inductive Probability

In his Essai philosophique sur les probabilités (1814), Laplace set out a mathematical system of inductive reasoning based on probability, which we would today recognise as Bayesian. He begins the text with a series of principles of probability, the first six being:

- Probability is the ratio of the "favored events" to the total possible events.
- The first principle assumes equal probabilities for all events. When this is not true, we must first determine the probabilities of each event. Then, the probability is the sum of the probabilities of all possible favored events.

- For independent events, the probability of the occurrence of all is the probability of each multiplied together.

- For events not independent, the probability of event B following event

A (or event A causing B) is the probability of A multiplied by the probability that A and B both occur.

- The probability that A will occur, given that B has occurred, is the probability of A and B occurring divided by the probability of B.

- Three corollaries are given for the sixth principle, which amount to Bayesian probability. Where event Ai ∈ {A1, A2, ...An} exhausts the list of possible causes for event B, Pr(B) = Pr(A1, A2, ..., An). Then

$$\Pr(A_i \mid B) = \Pr(A_i) \frac{\Pr(B \mid A_i)}{\sum_j \Pr(A_j) \Pr(B \mid A_j)}.$$

One well-known formula arising from his system is the rule of succession, given as principle seven. Suppose that some trial has only two possible outcomes, labeled "success" and "failure". Under the assumption that little or nothing is known a priori about the relative plausibilities of the outcomes, Laplace derived a formula for the probability that the next trial will be a success.

$$\Pr(\text{next outcome is success}) = \frac{s+1}{n+2}$$

where s is the number of previously observed successes and n is the total number of observed trials. It is still used as an estimator for the probability of an event if we know the event space, but have only a small number of samples.

The rule of succession has been subject to much criticism, partly due to the example which Laplace chose to illustrate it. He calculated that the probability that the sun will rise tomorrow, given that it has never failed to in the past, was

$$\Pr(\text{sun will rise tomorrow}) = \frac{d+1}{d+2}$$

where d is the number of times the sun has risen in the past. This result has been derided as absurd, and some authors have concluded that all applications of the Rule of Succession are absurd by extension. However, Laplace was fully aware of the absurdity of the result; immediately following the example, he wrote, "But this number [i.e., the probability that the sun will rise tomorrow] is far greater for him who, seeing in the totality of phenomena the principle regulating the days and seasons, realizes that nothing at the present moment can arrest the course of it."

Probability-generating Function

The method of estimating the ratio of the number of favorable cases to the whole number of possible cases had been previously indicated by Laplace in a paper written in 1779. It consists of treating the successive values of any function as the coefficients in the expansion of another function, with reference to a different variable. The latter is therefore called the probability-generating function of the former. Laplace then shows how, by means of interpolation, these coefficients may be determined from the generating function. Next he attacks the converse problem, and from the coefficients he finds the generating function; this is effected by the solution of a finite difference equation.

Least Squares and Central Limit Theorem

The fourth chapter of this treatise includes an exposition of the method of least squares, a remarkable testimony to Laplace's command over the processes of analysis. In 1805 Legendre had published the method of least squares, making no attempt to tie it to the theory of probability. In 1809 Gauss had derived the normal distribution from the principle that the arithmetic mean of observations gives the most probable value for the quantity measured; then, turning this argument back upon itself, he showed that, if the errors of observation are normally distributed, the least squares estimates give the most probable values for the coefficients in regression situations. These two works seem to have spurred Laplace to complete work toward a treatise on probability he had contemplated as early as 1783.

In two important papers in 1810 and 1811, Laplace first developed the characteristic function as a tool for large-sample theory and proved the first general central limit theorem. Then in a supplement to his 1810 paper written after he had seen Gauss's work, he showed that the central limit theorem provided a Bayesian justification for least squares: if one were combining observations, each one of which was itself the mean of a large number of independent observations, then the least squares estimates would not only maximize the likelihood function, considered as a posterior distribution, but also minimize the expected posterior error, all this without any assumption as to the error distribution or a circular appeal to the principle of the arithmetic mean. In 1811 Laplace took a different non-Bayesian tack. Considering a linear regression problem, he restricted his attention to linear unbiased estimators of the linear coefficients. After showing that members of this class were approximately normally distributed if the number of observations was large, he argued that least squares provided the "best" linear estimators. Here "best" in the sense that they minimized the asymptotic variance and thus both minimized the expected absolute value of the error, and maximized the probability that the estimate would lie in any symmetric interval about the unknown coefficient, no matter what the error distribution. His derivation included the joint limiting distribution of the least squares estimators of two parameters.

Laplace's Demon

In 1814, Laplace published what is usually known as the first articulation of causal or scientific determinism:

We may regard the present state of the universe as the effect of its past and the cause of its future. An intellect which at a certain moment would know all forces that set nature in motion, and all positions of all items of which nature is composed, if this intellect were also vast enough to submit these data to analysis, it would embrace in a single formula the movements of the greatest bodies of the universe and those of the tiniest atom; for such an intellect nothing would be uncertain and the future just like the past would be present before its eyes.

— Pierre Simon Laplace,
A Philosophical Essay on Probabilities

This intellect is often referred to as Laplace's demon (in the same vein as Maxwell's demon) and sometimes Laplace's Superman (after Hans

Reichenbach). Laplace, himself, did not use the word "demon", which was a later embellishment. As translated into English above, he simply referred to: "Une intelligence... Rien ne serait incertain pour elle, et l'avenir comme le passé, serait présent à ses yeux."

Even though Laplace is known as the first to express such ideas about causal determinism, his view is very similar to the one proposed by Boscovich as early as 1763 in his book Theoria philosophiae naturalis.

Laplace Transforms

As early as 1744, Euler, followed by Lagrange, had started looking for solutions of differential equations in the form:

$$z = \int X(x)e^{ax}\,dx \text{ and } z = \int X(x)x^{a}\,dx.$$

In 1785, Laplace took the key forward step in using integrals of this form in order to transform a whole difference equation, rather than simply as a form for the solution, and found that the transformed equation was easier to solve than the original.

Other Discoveries and Accomplishments

Mathematics

Amongst the other discoveries of Laplace in pure and applied mathematics are:

- Discussion, contemporaneously with Alexandre-Théophile Vandermonde, of the general theory of determinants, (1772);
- Proof that every equation of an even degree must have at least one real quadratic factor;
- Laplace's method for approximating integrals
- Solution of the linear partial differential equation of the second order;

- He was the first to consider the difficult problems involved in equations of mixed differences, and to prove that the solution of an equation in finite differences of the first degree and the second order might always be obtained in the form of a continued fraction; and

- In his theory of probabilities:

    - de Moivre–Laplace theorem that approximates binomial distribution with a normal distribution

    - Evaluation of several common definite integrals; and

    - General proof of the Lagrange reversion theorem.

Surface Tension

Laplace built upon the qualitative work of Thomas Young to develop the theory of capillary action and the Young–Laplace equation.

Speed of Sound

Laplace in 1816 was the first to point out that the speed of sound in air depends on the heat capacity ratio. Newton's original theory gave too low a value, because it does not take account of the adiabatic compression of the air which results in a local rise in temperature and pressure. Laplace's investigations in practical physics were confined to those carried on by him jointly with Lavoisier in the years 1782 to 1784 on the specific heat of various bodies.

- ROBERT STAWELL BALL

Politics

Minister of the Interior

In his early years Laplace was careful never to become involved in politics, or indeed in life outside the Académie des sciences. He prudently withdrew from Paris during the most violent part of the Revolution.

In November 1799, immediately after seizing power in the coup of 18 Brumaire, Napoleon appointed Laplace to the post of Minister of the Interior. The appointment, however, lasted only six weeks, after which Lucien, Napoleon's brother, was given the post. Evidently, once Napoleon's grip on power was secure, there was no need for a prestigious but inexperienced scientist in the government. Napoleon later (in his Mémoires de Sainte Hélène) wrote of Laplace's dismissal as follows:

Géomètre de premier rang, Laplace ne tarda pas à se montrer administrateur plus que médiocre; dès son premier travail nous reconnûmes que nous nous étions trompé. Laplace ne saisissait aucune question sous son véritable point de vue: il cherchait des subtilités partout, n'avait que des idées problématiques, et portait enfin l'esprit des 'infiniment petits' jusque dans l'administration.

(Geometrician of the first rank, Laplace was not long in showing himself a worse than average administrator; from his first actions in office we recognized our mistake. Laplace did not consider any question from the right angle: he sought subtleties everywhere, conceived only problems, and finally carried the spirit of "infinitesimals" into the administration.)

Grattan-Guinness, however, describes these remarks as "tendentious", since there seems to be no doubt that Laplace "was only appointed as a short-term figurehead, a place-holder while Napoleon consolidated power".

From Bonaparte to the Bourbons Laplace.

Pierre-Simon Laplace

Although Laplace was removed from office, it was desirable to retain his allegiance. He was accordingly raised to the senate, and to the third volume of the Mécanique céleste he prefixed a note that of all the truths therein contained the most precious to the author was the declaration he thus made of his devotion towards the peacemaker of Europe. In copies sold after the Bourbon Restoration this was struck out. (Pearson points out that the censor would not have allowed it anyway.) In 1814 it was evident that the empire was falling; Laplace hastened to tender his services to the Bourbons, and in 1817 during the Restoration he was rewarded with the title of marquis.

According to Rouse Ball, the contempt that his more honest colleagues felt for his conduct in the matter may be read in the pages of Paul Louis Courier. His knowledge was useful on the numerous scientific commissions on which he served, and, says Rouse Ball, probably accounts for the manner in which his political insincerity was overlooked.

Roger Hahn disputes this portrayal of Laplace as an opportunist and turncoat, pointing out that, like many in France, he had followed the debacle of Napoleon's Russian campaign with serious misgivings. The Laplaces, whose only daughter Sophie had died in childbirth in September 1813, were in fear for the safety of their son Émile, who was on the eastern front with the emperor. Napoleon had originally come to power promising stability, but it was clear that he had overextended himself, putting the nation at peril. It was at this point that Laplace's loyalty began to weaken. Although he still had easy access to Napoleon, his personal relations with the emperor cooled considerably. As a grieving father, he was particularly cut to the quick by Napoleon's insensitivity in an exchange related by Jean-Antoine Chaptal: "On his return from the rout in Leipzig, he [Napoleon] accosted Mr Laplace: 'Oh! I see that you have grown thin—Sire, I have lost my daughter—Oh! that's not a reason for losing weight. You are a mathematician; put this event in an equation, and you will find that it adds up to zero.'"

Political Philosophy

In the second edition (1814) of the Essai philosophique, Laplace added some revealing comments on politics and governance. Since it is, he says, "the practice of the eternal principles of reason, justice and humanity that produce and preserve societies, there is a great advantage to adhere to these principles, and a great inadvisability to deviate from them". Noting "the depths of misery into which peoples have been cast" when ambitious leaders disregard these principles, Laplace makes a veiled criticism of Napoleon's conduct: "Every time a great power intoxicated by the love of conquest aspires to universal domination, the sense of liberty among the unjustly threatened nations breeds a coalition to which it always succumbs." Laplace argues that "in the midst of the multiple causes that direct and restrain various states, natural limits" operate, within which it is "important for the stability as well as the prosperity of empires to remain". States that transgress these limits cannot avoid being "reverted" to them, "just as is the case when the waters of the seas whose floor has been lifted by violent tempests sink back to their level by the action of gravity".

About the political upheavals he had witnessed, Laplace formulated a set of principles derived from physics to favor evolutionary over revolutionary change:

> Let us apply to the political and moral sciences the method founded upon observation and calculation, which has served us so well in the natural sciences. Let us not offer fruitless and often injurious resistance to the inevitable benefits derived from the progress of enlightenment; but let us change our institutions and the usages that we have for a long time adopted only with extreme caution. We know from past experience the drawbacks they can cause, but we are unaware of the extent of ills that change may produce. In the face of this ignorance, the theory of probability instructs us to avoid all change, especially to avoid sudden changes which in the moral as well as the physical world never occur without a considerable loss of vital force.

In these lines, Laplace expressed the views he had arrived at after experiencing the Revolution and the Empire. He believed that the stability of nature, as revealed through scientific findings, provided the model that best helped to preserve the human species. "Such views," Hahn comments, "were also of a piece with his steadfast character."

Death

Laplace died in Paris in 1827. His brain was removed by his physician, François Magendie, and kept for many years, eventually being displayed in a roving anatomical museum in Britain. It was reportedly smaller than the average brain. Laplace was buried at Père Lachaise in Paris but in 1888 his remains were moved to Saint Julien de Mailloc in the canton of Orbec and reinterred on the family estate. The tomb is situated on a hill overlooking the village of St Julien de Mailloc, Normandy, France.

Tomb of Pierre-Simon Laplace

Posthumous portrait by
Jean-Baptiste Paulin Guérin, 1838.

Religious Opinions

I had no need of that hypothesis

A frequently cited but apocryphal interaction between Laplace and Napoleon purportedly concerns the existence of God. A typical version is provided by Rouse Ball:

Laplace went in state to Napoleon to present a copy of his work, and the following account of the interview is well authenticated, and so characteristic of all the parties concerned that I quote it in full. Someone had told Napoleon that the book contained no mention of the name of God; Napoleon, who was fond of putting embarrassing questions, received it with the remark, 'M. Laplace, they tell me you have written this large book on the system of the universe, and have never even mentioned its Creator.' Laplace, who, though the most supple of politicians, was as stiff as a martyr on every point of his philosophy, drew himself up and answered bluntly, Je n'avais pas besoin de cette hypothèse-là. ("I had no need of that hypothesis.") Napoleon, greatly amused, told this reply to Lagrange, who exclaimed, Ah! c'est une belle hypothèse; ça explique beaucoup de choses. ("Ah, it is a fine hypothesis; it explains many things.")

- ROBERT STAWELL BALL

An earlier report, although without the mention of Laplace's name, is found in Antommarchi's The Last Moments of Napoleon (1825):

Je m'entretenais avec L...... je le félicitais d'un ouvrage qu'il venait de publier et lui demandais comment le nom de Dieu, qui se reproduisait sans cesse sous la plume de Lagrange, ne s'était pas présenté une seule fois sous la sienne. C'est, me répondit-il, que je n'ai pas eu besoin de cette hypothèse. ("While speaking with L...... I congratulated him on a work which he had just published and asked him how the name of God, which appeared endlessly in the works of Lagrange, didn't occur even once in his. He replied that he had no need of that hypothesis.")

In 1884, however, the astronomer Hervé Faye affirmed that this account of Laplace's exchange with Napoleon presented a "strangely transformed" (étrangement transformée) or garbled version of what had actually happened. It was not God that Laplace had treated as a hypothesis, but merely his intervention at a determinate point:

In fact Laplace never said that. Here, I believe, is what truly happened. Newton, believing that the secular perturbations which he had sketched out in his theory would in the long run end up destroying the Solar System, says somewhere that God was obliged to intervene from time to time to remedy the evil and somehow keep the system working properly. This, however, was a pure supposition suggested to Newton by an incomplete view of the conditions of the stability of our little world. Science was not yet advanced enough at that time to bring these conditions into full view. But Laplace, who had discovered them by a deep analysis, would have replied to the First Consul that Newton had wrongly invoked the intervention of God to adjust from time to time the machine of the world (la machine du monde) and that he, Laplace, had no need of such an assumption. It was not God, therefore, that Laplace treated as a hypothesis, but his intervention in a certain place.

Laplace's younger colleague, the astronomer François Arago, who gave his eulogy before the French Academy in 1827, told Faye that the garbled version of Laplace's interaction with Napoleon was already in circulation towards the end of Laplace's life. Faye writes:

I have it on the authority of M. Arago that Laplace, warned shortly before his death that that anecdote was about to be published in a biographical collection, had requested him [Arago] to demand its deletion by the publisher. It was necessary to either explain or delete it, and the second way was the easiest. But, unfortunately, it was neither deleted nor explained.

The Swiss-American historian of mathematics Florian Cajori appears to have been unaware of Faye's research, but in 1893 he came to a similar conclusion. Stephen Hawking said in 1999, "I don't think that Laplace was claiming that God does not exist. It's just that he doesn't intervene, to break the laws of Science."

The only eyewitness account of Laplace's interaction with Napoleon is from the entry for August 8, 1802 in the diary of the British astronomer Sir William Herschel:

The first Consul then asked a few questions relating to Astronomy and the construction of the heavens to which I made such answers as seemed to give him great satisfaction. He also addressed himself to Mr Laplace on the same subject, and held a considerable argument with him in which he differed from that eminent mathematician. The difference was occasioned by an exclamation of the first Consul, who asked in a tone of exclamation or admiration (when we were speaking of the extent of the sidereal heavens): 'And who is the author of all this!' Mons. De la Place wished to shew that a chain of natural causes would account for the construction and preservation of the wonderful system. This the first

Consul rather opposed. Much may be said on the subject; by joining the arguments of both we shall be led to 'Nature and nature's God'.

Since this makes no mention of Laplace saying, "I had no need of that hypothesis," Daniel Johnson argues that "Laplace never used the words attributed to him." Arago's testimony, however, appears to imply that he did, only not in reference to the existence of God.

Views on God

Born a Catholic, Laplace appears for most of his life to have veered between deism (presumably his considered position, since it is the only one found in his writings) and atheism.

Faye thought that Laplace "did not profess atheism", but Napoleon, on Saint Helena, told General Gaspard Gourgaud, "I often asked Laplace what he thought of God. He owned that he was an atheist." Roger Hahn, in his biography of Laplace, mentions a dinner party at which "the geologist Jean-Étienne Guettard was staggered by Laplace's bold denunciation of the existence of God". It appeared to Guettard that Laplace's atheism "was supported by a thoroughgoing materialism". But the chemist Jean-Baptiste Dumas, who knew Laplace well in the 1820s, wrote that Laplace "provided materialists with their specious arguments, without sharing their convictions".

Hahn states: "Nowhere in his writings, either public or private, does Laplace deny God's existence." Expressions occur in his private letters that appear inconsistent with atheism. On 17 June 1809, for instance, he wrote to his son, "Je prie Dieu qu'il veille sur tes jours. Aie-Le toujours présent à ta pensée, ainsi que ton père et ta mère [I pray that God watches over your days. Let Him be always present to your mind, as also your father and your mother]." Ian S. Glass, quoting Herschel's account of the celebrated exchange with Napoleon, writes that Laplace was "evidently a deist like Herschel".

In Exposition du système du monde, Laplace quotes Newton's assertion that "the wondrous disposition of the Sun, the planets and the comets, can only be the work of an all-powerful and intelligent Being". This, says Laplace, is a "thought in which he [Newton] would be even more confirmed, if he had known what we have shown, namely that the conditions of the arrangement of the planets and their satellites are precisely those which ensure its stability". By showing that the "remarkable" arrangement of the planets could be entirely explained by the laws of motion, Laplace had eliminated the need for the "supreme intelligence" to intervene, as Newton had "made" it do. Laplace cites with approval Leibniz's criticism of Newton's invocation of divine intervention to restore order to the Solar System: "This is to have very narrow ideas about the wisdom and the power of God." He evidently shared Leibniz's astonishment at Newton's belief "that God has made his machine so badly that unless he affects it by some extraordinary means, the watch will very soon cease to go".

- ROBERT STAWELL BALL

In a group of manuscripts, preserved in relative secrecy in a black envelope in the library of the Académie des sciences and published for the first time by Hahn, Laplace mounted a deist critique of Christianity. It is, he writes, the "first and most infallible of principles ... to reject miraculous facts as untrue". As for the doctrine of transubstantiation, it "offends at the same time reason, experience, the testimony of all our senses, the eternal laws of nature, and the sublime ideas that we ought to form of the Supreme Being". It is the sheerest absurdity to suppose that "the sovereign lawgiver of the universe would suspend the laws that he has established, and which he seems to have maintained invariably".

In old age, Laplace remained curious about the question of God and frequently discussed Christianity with the Swiss astronomer Jean-Frédéric-Théodore Maurice. He told Maurice that "Christianity is quite a beautiful thing" and praised its civilizing influence. Maurice thought that the basis of Laplace's beliefs was, little by little, being modified, but that he held fast to his conviction that the invariability of the laws of nature did not permit of supernatural events. After Laplace's death, Poisson told Maurice, "You know that I do not share your [religious] opinions, but my conscience forces me to recount something that will surely please you." When Poisson had complimented Laplace about his "brilliant discoveries", the dying man had fixed him with a pensive look and replied, "Ah! we chase after phantoms [chimères]." These were his last words, interpreted by Maurice as a realization of the ultimate "vanity" of earthly pursuits. Laplace received the last rites from the curé of the Missions Étrangères (in whose parish he was to be buried) and the curé of Arcueil.

However, according to his biographer, Roger Hahn, since it is "not credible" that Laplace "had a proper Catholic end", the "last rights" (sic) were ineffective and he "remained a skeptic" to the very end of his life. Laplace in his last years has been described as an agnostic.

Excommunication of a Comet

In 1470 the humanist scholar Bartolomeo Platina wrote that Pope Callixtus III had asked for prayers for deliverance from the Turks during a 1456 appearance of Halley's Comet. Platina's account does not accord with Church records, which do not mention the comet. Laplace is alleged to have embellished the story by claiming the Pope had "excommunicated" Halley's comet. What Laplace actually said, in Exposition du système du monde (1796), was that the Pope had ordered the comet to be "exorcised" (conjuré). It was Arago, in Des Comètes en général (1832), who first spoke of an excommunication.

Honors

- Correspondent of the Royal Institute of the Netherlands in 1809.
- Foreign Honorary Member of the American Academy of Arts and Sciences in 1822.
- The asteroid 4628 Laplace is named for Laplace.
- A spur of the Montes Jura on the moon is known as Promontorium Laplace.
- His name is one of the 72 names inscribed on the Eiffel Tower.
- The tentative working name of the European Space Agency Europa Jupiter System Mission is the "Laplace" space probe.
- A train station in the RER B in Arcueil bears his name.

Quotations

- I had no need of that hypothesis. ("Je n'avais pas besoin de cette hypothèse-là", allegedly as a reply to Napoleon, who had asked why he hadn't mentioned God in his book on astronomy.)

- It is therefore obvious that ... (Frequently used in the Celestial Mechanics when he had proved something and mislaid the proof, or found it clumsy. Notorious as a signal for something true, but hard to prove.)

- "We are so far from knowing all the agents of nature and their diverse modes of action that it would not be philosophical to deny phenomena solely because they are inexplicable in the actual state of our knowledge. But we ought to examine them with an attention all the more scrupulous as it appears more difficult to admit them."

- This is restated in Theodore Flournoy's work From India to the Planet Mars as the Principle of Laplace or, "The weight of the evidence should be proportioned to the strangeness of the facts."

- Most often repeated as "The weight of evidence for an extraordinary claim must be proportioned to its strangeness."

- This simplicity of ratios will not appear astonishing if we consider that all the effects of nature are only mathematical results of a small number of immutable laws.

- Infinitely varied in her effects, nature is only simple in her causes.

- What we know is little, and what we are ignorant of is immense. (Fourier comments: "This was at least the meaning of his last words, which were articulated with difficulty.")

ABOUT THE AUTHOR

ROBERT STAWELL BALL
(July 1, 1840 – November 25, 1913)

Sir Robert Stawell Ball was an Irish astronomer who founded the screw theory.

He was the son of naturalist Robert Ball and Amelia Gresley Hellicar. He was born in Dublin.

Ball worked for Lord Rosse from 1865 to 1867. In 1867 he became Professor of Applied Mathematics at the Royal College of Science in Dublin. There he lectured on mechanics and published an elementary account of the science.

In 1874 Ball was appointed Royal Astronomer of Ireland and Andrews Professor of Astronomy in the University of Dublin at Dunsink Observatory.

Ball contributed to the science of kinematics by delineating the screw displacement:

When Ball and the screw theorists speak of screws they no longer mean actual cylindrical objects with helical threads cut into them but the possible motion of any body whatsoever, including that of the screw independently of the nut.

Ball's treatise *The Theory of Screws* (1876) is now in the public domain. His work on screw dynamics earned him in 1879 the

Cunningham Medal of the Royal Irish Academy.

In 1882 Popular Science Monthly carried his article "*A Glimpse through the Corridors of Time*". The following year it carried his two-part article on "*The Boundaries of Astronomy*".

Ball expounded the tides in *Time and Tide*: a Romance of the Moon In 1892 he was appointed Lowndean Professor of Astronomy and Geometry at Cambridge University at the same time becoming director of the Cambridge Observatory. He was a fellow of King's College, Cambridge.

In 1900 Cambridge University Press published *A Treatise on the Theory of Screws*. That year he also published *The Story of the Heavens*. Much in the limelight, he stood as President of the Quaternion Society. He was also President of the Mathematical Association in 1900.

In 1908 he published *A Treatise on Spherical Astronomy*, which is a textbook on astronomy starting from spherical trigonometry and the celestial sphere, considering atmospheric refraction and aberration of light, and introducing basic use of a generalised instrument.

His work *The Story of the Heavens* is mentioned in the "Ithaka" chapter of Ulysses. His lectures, articles and books (e.g. *Starland* and *The Story of the Heavens*) were mostly popular and simple in style.

He died in Cambridge and was buried at the Parish of the Ascension Burial Ground in Cambridge, with his wife Lady Francis Elizabeth Ball. Their children were: Frances Amelia, Robert Steele, William Valentine (later Sir), Mary Agnetta, Charles Rowan Hamilton, and Randall Gresley (later Colonel).

LECTURES

Ball became celebrated for his popular lectures on science. He gave an estimated 2500 lectures between 1875 and 1910 in towns and cities across Britain and Ireland. In 1892, 1898 and 1900 he was invited to deliver the Royal Institution Christmas Lecture. *Astronomy; Astronomy* and *Great Chapters from the Book of Nature.*

FEEDBACK

Now that you have read the book ...

Was it interesting?

Did you enjoy what you wanted to read?
Was there any room for improvement?

Let us know at:

http://www.diamondbooks.ca/feedback

Your feedback is highly appreciated.

Thank you!

- ROBERT STAWELL BALL

NOTES

NOTES

NOTES

NOTES

NOTES

NOTES

Made in United States
Cleveland, OH
06 November 2026

47647261R00072